The Question & Answer Book

ALL ABOUT ISLANDS

ALL ABOUT ISLANDS

By Wendy Rydell
Illustrated by Ray Burns

Troll Associates

Library of Congress Cataloging in Publication Data

Rydell, Wendy.
    All about islands.

    (Question and answer book)
    Summary: Answers questions about the formation of
islands by natural forces and about the development of
distinctive plant and animal life on individual islands.
        1. Islands—Juvenile literature.   [1. Islands.
2. Island ecology.   3. Ecology.   4. Questions and
answers]   I. Burns, Raymond, 1924-   ill.   II. Title.
III. Series.
GB461.R9   1984   551.4'2   83-4833
ISBN 0-89375-975-9
ISBN 0-89375-976-7 (pbk.)

How is an island born?

Many, many years ago, at a spot deep beneath a great ocean, a small crack opened in the earth's hard surface.

Fire-hot molten rock from the earth's liquid center pushed up through the crack, widening it just a little more.

When the molten rock, called lava, touched the water of the ocean floor, it sizzled, cooled, and hardened into solid rock. But the crack stayed open. More lava, forced upward by great pressures in the earth's fiery core, came pouring out through it.

5

Over countless years, the flow of lava continued. Each time the hot lava pushed up, it cooled and hardened, forming a mound around the crack in the ocean floor. With each new burst of lava, the mound grew higher and higher, until it reached halfway to the surface of the ocean.

The mound of hardened rock had become a mountain—a roaring, spitting, underwater volcanic mountain. It kept growing taller and taller with each new burst of lava. More time went by. The volcano kept right on growing, although its great size remained hidden under the ocean.

What is a river island?

The special one-of-a-kind world of an island may be an *inland island*. Inland islands are surrounded by water, and that water is surrounded by land. For example, some inland islands are *river islands*. A river island may form when hard earth forces a river to flow around it, leaving the land firm and dry, but surrounded by water. Or a river island may be built up over the years by dirt and sand that are swept along by the river's waters and then dropped at a spot where the riverbed is very shallow.

What is a lake island?

Some inland islands are *lake islands*. A lake island may have been pushed up out of the lake floor by the thrust of a powerful earthquake. Or it may have been left standing when a great Ice Age glacier pushed away the surrounding earth thousands of years ago. Then the glacier melted to form a lake.

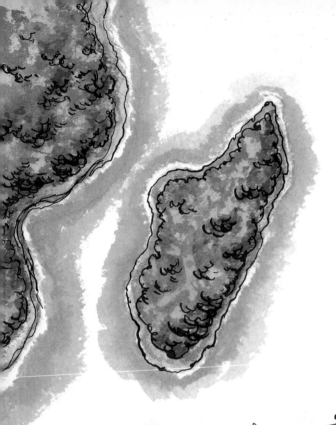

What is a coastal island?

Continental or *coastal islands* lie near the coasts of the large land masses called continents. These islands may be slices of land that were cut off from the mainland by earthquakes or by the constant pounding of wind and waves. They may have risen up from the continental shelf—the underwater land surrounding most continents. Or they may be bars of sand and silt, built by the shifting tides that often swirl around many coastal beaches.

What is an oceanic island?

Most of the world's islands lie far from any other land. They are called *oceanic islands*. An oceanic island may be the peak of a volcano that rises up from the ocean floor. Or it may be the rock-hard skeletons of countless tiny sea creatures called corals.

12

How do corals build islands?

There are tiny corals called reef-building corals, or *polyps*. Each one is a living island builder. It produces a substance called lime, which forms a hard, skeleton-like protection around the polyp's body. Reef-building corals attach themselves like a fringe to rocks in the shallow water near the shores of a volcanic island.

When the polyps die, their hard skeletons remain attached to the underwater rocks. New polyps attach themselves to the skeletons. When they die, the coral fringe gets thicker. This kind of coral formation is known as a *fringing reef*.

What is a barrier reef?

Sometimes, a volcanic island begins sinking toward the ocean floor. As the land sinks, the coral formation continues to build upward. Now the coral reef is separated from the island by a wide stretch of calm water. The reef forms a barrier between the ocean and the calm water around the island. It is called a *barrier reef.*

What is an atoll?

As time goes on, an island surrounded by a barrier reef may disappear completely under the sea. The barrier reef continues to grow. Now it forms a ring around a quiet lagoon. This kind of coral formation is called an *atoll.*

Yet, the forces of nature never stand still. Even while new generations of coral polyps keep building the atoll, the ocean's powerful waves keep wearing it down. Eventually, they will divide the atoll into a series of separate coral islands. One day, even these little islands—which were formed by living island builders—may disappear beneath the surface of the sea.

Can plants build islands?

Yes. One kind of living island builder is an amazing "walking" tree called the mangrove. Mangroves usually grow in warm climates where rivers empty into tropical seas.

Mangroves begin as tiny seedlings that take root in the warm, shallow water. But very quickly they grow several leafy branches that send out roots of their own. These new roots, which are called prop roots, reach down from the spreading branches.

In just a few years, hundreds of prop roots have spread out in all directions. Each one grows thicker and stronger. Each one reaches out farther into the shallow water, as if the mangrove tree were "walking."

Silt, shells, and bits of floating sea grass are trapped by the thick, spreading roots. And as time goes on, widening patches of fertile soil are built up. Other mangrove seeds sprout in this soil, and they, too, grow and spread. The trees and the built-up soil around them become a mangrove island.

What lives on an island?

No matter how it was formed, an island is usually the home of living things.

A river island often has the same kinds of flowers and plants that grow on the nearby river banks. Continental or coastal islands are fairly close to land, so they also usually have the same plants, insects, and animals that live on the mainland. But this is not always so.

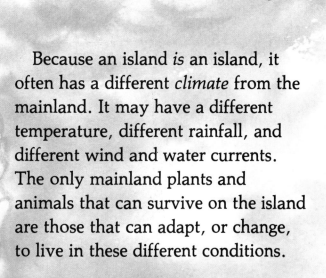

Because an island *is* an island, it often has a different *climate* from the mainland. It may have a different temperature, different rainfall, and different wind and water currents. The only mainland plants and animals that can survive on the island are those that can adapt, or change, to live in these different conditions.

17

How does climate affect the Scilly Islands?

The Scilly Islands near England are right in the path of a warm current of ocean water. The current heats the air above the islands and gives them an almost tropical climate. As a result, palm trees, bananas, and other tropical plants thrive on these unusual islands. But the same plants could never survive on the colder island of England, only 25 miles (40 kilometers) away.

How does size affect an island?

Sometimes the size of an island helps to decide what will live on it. Small animals, such as field mice, can easily survive on a small island. But larger animals may not find enough to eat. So some islands become *sanctuaries*, or safe places, for certain kinds of animals.

Where are some island sanctuaries?

There are many island sanctuaries in the world today. On the Pribilof Islands in the Bering Sea and on the island of Guadeloupe in the West Indies, thousands of fur seals come to mate. On Ascension Island in the South Atlantic, green sea turtles lay their eggs. On the tiny Scottish island of Saint Kilda, millions of sea birds come to nest.

How do plants and animals get to an island?

Plants and animals can easily float, fly, or swim from the mainland to inland and coastal islands. But how do plants and animals reach islands that are far from other bodies of land?

Oceanic islands are surrounded by vast stretches of rolling sea. Yet even these isolated places are filled with life. Often they are the most interesting of the world's living laboratories. Let's visit a newborn volcanic island and find out how life reaches it.

For a long time after its fiery birth, the volcanic island was not much more than a tiny dot of rock in the middle of the ocean. Thunderous waves nearly washed away each new layer of lava as soon as it came spitting out of the volcano's cone. But some of the lava *did* remain. It cooled and hardened into black volcanic rock.

Now another powerful force—the weather—began changing the island's rock-hard face. Heated by the sun and cooled by the rain, the surface rocks split and crumbled. As time went on, the pieces of rock were ground into a layer of sand. Blown there by the wind, dust settled on the island, and ocean storms washed debris ashore. In time, there was a layer of soft, crumbly soil.

Nothing lived on the island. Nothing grew in its new soil. But that would soon change.

What came first?

One day, a sea gull was blown far from its usual course by a storm. It landed on the island and waited for the storm to pass. When the gull went, it left behind a few seeds that had been entangled in its feathers. The seeds took root in the soil and sprouted.

Other seeds reached the island in other ways. Some were carried by air currents from distant islands. Some were carried by driftwood that bobbed along in the ocean until it was washed up on the beach. In time, bits of green appeared on the island. There were patches of grass and ferns and other plants whose seeds floated easily on the wind and water.

Some of these first plants died. But as they decayed, they made the soil richer. And soon, new seeds sprouted and grew, until much of the island was covered with a green carpet of plants. The island was ready for its first animal life.

Which animals came first?

One day a curled-up palm leaf washed onto the beach. The leaf soon rotted away in the sun, but not before its passengers—a colony of ants and a tiny spider with an egg sac attached to its body—crawled away to safety. Unlike the sea gull, these tiny creatures stayed. And they thrived, because now there were tasty leaves and roots and seeds to eat.

23

Year in and year out, the wind and waves brought more creatures to the island. A lizard arrived on a floating coconut shell. Butterflies flew from a distant island, and a swarm of dragonflies came from another.

A raging storm brought a group of land snails and beetles, sheltered inside the branches of an uprooted mango tree. After the storm, some of the mango seeds sprouted, and the island's first trees took root. Flocks of birds, blown to the island by the wind, built their nests in the trees.

24

What happened next?

As the years went by, the island developed its own special mixture of living things. There were birds, reptiles, sea animals, insects, shrubs, and trees.

Some plants, no longer needing to share soil and water with other mainland plants, grew amazingly large. Other plants, twisted by the island's constant winds, took on new and beautiful shapes.

Some reptiles became smaller and smaller with each new generation, because they had difficulty finding enough food. Others grew bigger and bigger, because their natural enemies were far away on the mainland.

Some birds no longer needed strong wings to fly around the tiny island. In time, the wings of this type of bird became smaller and weaker. At the same time, their legs grew bigger and stronger, because they used them more.

Some animals, separated by the ocean from the rest of their kind, mated with slightly different animals on the island. In time, they produced a new and different kind of animal—a species found only on this one island.

Of all the plants and animals that reached the island by chance, the ones that survived—and increased in number —were the ones best able to adapt to island conditions.

What are some unusual island creatures?

Today, there are many islands where strange or unusual plants and animals have developed. The islands of Indonesia are the home of a rare creature called the Komodo dragon. This giant meat-eating lizard grows to a length of almost 10 feet (3 meters) and weighs as much as 300 pounds (135 kilograms). Its mainland relatives rarely grow to even half that size.

The tuatara, a strange lizard-like creature whose nearest relatives died out more than a million years ago, still survives on a few tiny islands off the coast of New Zealand.

28

The Galapagos Islands in the Pacific Ocean are perhaps nature's finest living laboratories. They have produced many unusual creatures, including the world's largest land tortoises and the world's only marine iguanas.

On the island continent of Australia, large numbers of marsupials have developed and survived. Marsupials are mammals that carry their young in pouches.

Insects and birds that have lost the ability to fly have developed on a number of islands. The wingless beetles of the Galapagos and the kiwi birds of New Zealand are two examples. The flightless dodo bird developed on the island of Mauritius in the Indian Ocean. It survived until about 300 years ago, when hunters killed off the last of these strange birds.

As people discovered, explored, and settled on islands, they often upset the delicate balance of nature. Sometimes people hunted animals that other creatures needed in order to survive. Sometimes their dogs, cats, goats, or pigs preyed on island creatures that had no natural enemies.

What can be done to protect our islands?

Today, more and more people are beginning to understand just how special our islands and their unique plants and animals are. As a result, parts of some islands have been preserved as national parks and wildlife sanctuaries. Now, many island creatures are being saved by laws that protect them.

Meanwhile, new islands are constantly being formed by the forces of nature. In time, they, too, will become special little worlds of their own—living laboratories with their own special and fascinating mixture of plants and animals.